THE FATE OF MY FATHER

THE TRIUMPHANT TRUE STORY OF A FATHER WHO CHALLENGED POWER AT HOME AND OVERSEAS

WARRICK L. BARRETT

JONATHAN GREEN

DRAGON GOD BOOKS

Copyright © 2017-2018 by Dragon God, Inc.

All rights reserved.

Simultaneously published in United States of America, the UK, India, Germany, France, Italy, Canada, Japan, Spain, and Brazil.

All rights reserved. No part of this book may be reproduced in any form or by any other electronic or mechanical means – except in the case of brief quotations embedded in articles or reviews –without written permission from its author.

The Fate of my Father has provided the most accurate information possible. Many of the techniques used in this book are from personal experiences. The author shall not be held liable for any damages resulting from use of this book.

Paperback ISBN-13: 978-1977558824

Paperback ISBN-10: 1977558828

Hardback ISBN: 978-1947667136

CONTENTS

1. Taking Flight — 1
2. Childhood — 3
3. High School — 8
4. Wilberforce — 13
5. College Football — 16
6. Fort Sill — 21
7. West Virginia — 24
8. Officer Candidate School — 33
9. World War II — 37
10. Semi-Professional Football — 44
11. Military Career — 46
12. Marriage — 52
13. Final Military — 58
14. Education Part II — 62
15. High School Leadership — 64
16. Civil Air Patrol — 68
17. Organizations — 72
18. Following His Footsteps — 74
19. Saying Goodbye — 78

Found a Typo? — 81
Books by Jonathan Green — 83
One Last Thing — 85

1

TAKING FLIGHT

If there was one thing Ross Paige Barrett knew how to do, that was making an entrance. That day, he showed that he also knew a thing or two about making an exit. He took off that morning never knowing that it would be his last day on this planet.

Every pilot has to recertify periodically because piloting is a very dangerous profession – not only for the pilot, but also for everyone else on the ground. There are many federal civil aviation regulations to ensure that pilots are competent and healthy enough to fly. My father was not a young man, and he had completed his aviation certification exam many times, but the flight instructor administering my father's test had no idea how that day would turn out.

On 3rd January 1985, as he took to the skies with his aviation examiner in the co-pilot seat next to him, that man sat there with a clipboard in his hand, checking off as my father went through the proper procedures for takeoff, rating, cruising, and altitude. He asked my father to perform various tasks to demonstrate that he knew how to turn left, right, and how to deal with different situations. As he was looking down, he gave my father one final task, but nothing happened. The man repeated himself, asking my father again, "Sir, you need to complete this maneuver if you want to maintain your certification." The third time, he looked over and realized my father was gone.

Like a true hero, my father waited until he was halfway to heaven to leave his body behind – what better way to depart this world than soaring high with the world beneath your feet? A black man who had dreamed of being a Tuskegee Airman, stepping out of this world into the next one halfway to the sky. By the look my father's flight examiner had on his face, this must have been an experience that he would never forget.

That is the story of my father, a man you would probably never forget once you met him. Nearly thirty years later, I am sure that same aviation examiner still occasionally wakes up at night soaked in sweat, remembering the time this pilot left him behind – what better way to leave this world than surprising and even scaring the people around you.

My father's entire life was leading up towards this point. He was not someone who ever peaked at any point in his life; as one career ended, so the next one began. He never gave up, and he never stopped soaring higher.

In this book is the story not only of his journey, but also my journey as his son. As I clung to his shoulders, soaring on the back of an eagle, my journey is nothing but the continuation of his, as I live in a world that my father paid a critical role in creating.

2
CHILDHOOD

My father once told my mother that he felt he had been born twenty years too early. If he had been born just twenty years later, he felt he could have accomplished everything he dreamed of in this life. He was born into a world where not all men were treated equally. He was born into a world that was about to go to war for the second time. He was born into a world where opportunities were limited for a black man.

Even as a child, he was never someone who accepted everyone else's rules. He knew he could accomplish amazing things, and he pushed his life forward to change the world one step at a time.

My father was a natural athlete, and as he grew up, he only became stronger. If it had not been for World War II and segregation within the National Football League (NFL), my father very well could have played professional football. Later in his career, even after fighting for his country, he managed to play semi-professional football.

This story begins with the childhood of a man who grew up in a simple family in the Midwest in Cambridge, Ohio. He went to Park Elementary School at a time when big city schools were often still segregated, and you had to fight and compete for better education. In

Cambridge, the student bodies had a small number of minority students, mixed in with a white majority.

As my father grew from a boy into a young man, he believed in greatness. He had big dreams, and when you have big dreams people around you will often tell you that they are never going to come true. People will tell you that you are never going to fly an airplane, you are never to be a hero, and you will never be an officer in the military because you do not have what it takes – you are not tall enough, you are not smart enough, you are not strong enough. My father is an example that when you believe enough and ignore the people around you and all the naysayers who hold you back, you can accomplish amazing things.

His passion for education was circular; he loved learning and later in life he loved teaching. He was a man who believed in giving back and passing on knowledge; he believed that children stood on their fathers' shoulders and that one day their children would stand on their shoulders. I tell my father's story because I have an obligation to continue living his dream, to continue sharing his passion, and to continue affecting the world through the power of his journey.

The world is an ever-changing place. The building where my father went to elementary school is long gone; in 1952, it was replaced by a new, better building, and that building is gone too. We live in a world of constant change and we can either accept those changes, accept a life of entropy or watch the world collapse around us, or we can fight for a better world.

1322 Wheeling Avenue - Cambridge Ohio - the home Dad grew up in

The Barretts, Cambridge, OH. circa 1932. Top, L to R: Ross Barrett, Leauvinia Barrett, Nellie Barrett, Lee Barrett; below: Clara Ann Barrett

Park Elementary School

Park Elementary 2-3 grade

3
HIGH SCHOOL

In high school, life became even more exciting. At the time, education and sports were often segregated in big cities; thoughts of integration were on many people's minds, but my father did not need to wait for that. In Cambridge, he had the opportunity to enjoy integration as a child. He believed that if you work hard enough, you can demonstrate through your effort and your character what you are truly capable of.

As a high school football player, his skills earned him acclaim. He was good enough that the *Pittsburgh Courier* made him one of their high school All-American football players (for a long time this newspaper was an institution; it was the newspaper for black people in America). He earned first team all-league honors. He felt that if his Cambridge Brown High School team had enjoyed a more successful season, he might have earned all-state honors as well. At the time, you could be a star, but you were still only limited to a separate world. There were two Americas back then, and by the time my father died, he had lived in both.

As sons and daughters, we often rebel against our parents. We go through a phase, especially during our teenage years, where we desire to become our own people and to escape them. As we get older, we realize that we simply become new versions of our parents. It is true that some parents live out their dreams in their children, and they become too involved in their lives, but if it weren't for my father's journey, my high school journey would have been completely different.

I followed his footsteps as a high school athlete myself. He had a dream that his son could go one step further to play college football and maybe even break into the NFL. My father simply wanted me to achieve the things that he was so close to achieving.

The Cambridge *Daily Jeffersonian* described my father as a "rangy lad" in one of their preseason articles about his football prowess. These days there is a lot of debate about whether or not college is worth it and whether it leads you to a better career, but back then, getting into a college was tough. It was not an option for many people, and it certainly would not have been an option for my father, if he had not used his football skills to separate himself from the crowd. When an obstacle is before you, you have two options: you can see it as a barrier, or you can see it as an opportunity. He knew that if he could overcome that single barrier, the competition would thin out, and he was right.

As a proud football player, he was nearly unstoppable, and he was quite often victorious. I am proud to say that I followed in my father's footsteps, and while I was not that good in high school, I managed to step up in college and make a real difference for my team. I was not quite good enough to make the NFL, but I did take my career a few steps further than my father, and I will never forget the look of pride I saw in his eyes when I took the field and excelled in a predominantly white college.

In 1940, just as World War II was beginning to grow into a true monster in Europe, my father graduated from high school. As a brilliant athlete and a scholar, he was one of those students who would put his head down and do his best, and that true effort won him a scholarship to the Wilberforce University.

At the time, Wilberforce's football team was led by the renowned head coach Gaston "Country" Lewis. The Bulldogs were one of the most feared black college football teams of their generation, and other schools called them the "Force Eleven" – they were that afraid of my father's squad. Only the greatest of football teams get additional nicknames. It might not have been like the "Steel Curtain" of the 1979 Pittsburgh Steelers, but I reckon having a nickname for your college football team is pretty good.

He played with some amazing athletes, and if it had not been for the intervention of World War II, it is very possible my father could have been one of those men the NFL integrated just a couple of years after his graduation date.

Sometimes life gets in the way. As part of my father's academic decisions, he joined the Reserve Officer Training Corps (ROTC) program, training to be part of the military, because he saw something. He was always thinking three steps ahead, like a master chess player. While he hoped that he would be allowed to take his football career to the next level, he saw that it was not allowed quite yet. He was a few years ahead of the curve, and I am proud of him for that. He was ready for that opportunity when it happened, and he saw that in the military a man was judged by nothing other than his skills, his ferocity in combat, his bravery against adversity, and the way he could lead men.

He believed in his country, and he saw that an education at one of the premier institutions in the United States combined with strong military career would open up many opportunities. Alas, he never graduated from Wilberforce. His ROTC unit was called up, his ticket was selected, and my father was called to Europe to fight and bleed for his country – the country that would not yet admit that he was just as good as everyone else.

The Fate Of My Father

ROSS PAIGE BARRETT — A nicer boy you wouldn't want to meet. He is very generous and kind. He has done a fine job in our football games.

Cambridge High School 'Cantab' yearbook, 1940

Cambridge High School Alumni honor

Cambridge High School Football Team - Fall 1939 (Middle Back Row)

4

WILBERFORCE

One of the professors at the illustrious Wilberforce University was E. Champ Warrick. He taught a variety of liberal arts subjects and was the professor of secondary education. The influence of this man's wisdom on my father's life would be felt for another fifty years; he planted the seeds that led to my father becoming a fabulous, well-respected and loved secondary education teacher.

Before we get too far ahead of ourselves, there is another part of our story that is quite critical. Like many college professors, Prof. Warrick had a daughter, the very attractive Allyson Lightfoot Warrick. They both came into Wilberforce University in the fall of 1940. The school was under the administration of the AME (African Methodist Episcopal) church. My father and his family had belonged to the Shaffer Chapel AME church in Cambridge. At Wilberforce University, there were church gatherings that were pretty much mandatory for all the students, and I guess it was at one of those meetings that my father first met my mother.

When my father first set eyes upon this young, beautiful debutante, he felt a pitter patter in his heart. She was very pretty, and he was smitten with her beauty much right off the bat. For their first date, Dad asked her to go to the movies. From what I can remember, the thing that attracted my mother to my father was the genuine, nice, giving, kind and uncomplicated person that he was.

Mom was from the little village of Wilberforce, where Wilberforce University was, so she wasn't a "city slicker" type of woman. She was attracted to a more down-to-earth, plain and simple, good-looking guy, whom she had also probably heard that he was one of the stars on the football team. Their relationship took off from there.

They began spending time together, as college students do, and one thing led to another. There was a feeling in my father's heart that he wanted to spend the rest of his life with this woman, but when your country comes calling, sometimes you have to make tough decisions. You can't say, "Sorry Mr. President, I can't go to war right now; I am just beginning to fall in love with someone." It was too soon in the relationship for my father to get down on one knee, but he did not want to lose the woman of his dreams.

Their courtship went on between the fall of 1940, when they met, until sometime in 1943, when the Advanced Course Cadets of the Army ROTC program got activated for the Second World War. That, of course, caused separation between the two, sometime in 1943, when Dad had to pull away from school going off of his officer military training and ultimately went from that to war. As he was getting ready to be sent off to fight for his country and risk everything, he asked her, "I know that the second I am gone every other guy around here is going to have his eyes on you, but I ask that you wait for me." He knew that this beautiful and intelligent young woman had her choice of men in the world, and he could only go to war praying she wouldn't forget him. He pretty much had a fixation on wanting my mother to be his future bride. An ongoing theme in a lot of letters was "please wait for me."

I remember my mother telling me that her father, my Grandpa Warrick, who was a member of the Wilberforce faculty, for some reason was not overly impressed with my Dad. Maybe he was looking for someone sharp and more sophisticated. Whereas my mother's mother, Grandma Warrick, just fell in love with my father right away.

My grandmother Warrick had grown up in Wheeling, West Virginia, which was not too far from Cambridge Ohio, where my father grew up. I think it is maybe 80 miles or thereabouts between the Wheeling, West Virginia, and Cambridge, Ohio. My father even happened to grow up on Wheeling Avenue – specifically 1322 East Wheeling Avenue, in Cambridge, Ohio.

My mother was pretty impressed with my Dad, so she was willing to wait until after the war. She rejected a lot of courtship, guys from the university, and those city slicker guys. Wilberforce was a rural area, but there were a lot of big-city guys coming there, including from Dayton, Detroit, Cincinnati, Columbus, Cleveland, New York, and Washington DC. All of the more sophisticated, fast-moving, the moving and grooving kind of guys.

Just after Dad left to go to the military in 1943, she finished her college degree at the Wilberforce University class of 1944. I think her first job out of college was a teaching position in Detroit. She was in a rooming house run by a lady that was introduced me as "Mother Gregory." Her given name was Winifred Gregory. Mom stayed with Mother Gregory for a couple of years between 1944 and 1946. I guess she just did whatever a young teacher might do in Detroit – she was probably hanging around with the ladies, there were several Wilberforce the alumni, probably some classmates of hers who were in Detroit. That is probably who she hung out with while under the guidance and tutorship of Mother Gregory. How to be a woman, how to take care of yourself, that kind of thing. I recall that one of her best friends from Detroit was Jean Lane, who later became Jean Lane Parker.

5

COLLEGE FOOTBALL

My father was number thirty-six on the Wilberforce football team. I am lucky enough that a few photos of him with his team have survived all the way from back then to now, and I am excited to share them with you in this book. As you can see, my father was a very big and very strong man. Because he was so conservative and a bit unsophisticated, the players and other fellow students began to call him by the nickname "Lil' Abner." If you do not recall, Lil' Abner was a comic strip from way back then, about a big strong ox of a country bumpkin. Even though this comic strip character was white, my father was similar enough to him he was locked into that nickname throughout his short but intense college football career.

He only played at Wilberforce for a short time before being called up to the military, but my father and a few of the top Wilberforce linemen squad mates made the first page of the sports section of November 21, 1941 edition of the *Pittsburgh Courier*, which was published before a big game between "The Force" and West Virginia State played at the University of Detroit's stadium. They played in many games to be victorious – it was an exciting time. One of his closest friends of that time was Clifton "Ghost" Brown from Lebanon,

Ohio who in the 1970s went on to become the mayor of Marysville, Ohio and who helped encourage the Honda corporation to establish a large facility there. Ironically, during his time playing for Wilberforce side-by-side with my father, he was nicknamed "The Mayor." Another of his good friends and teammates was Charles "Reverend" Walton of Birmingham, Alabama. Interestingly enough, Charles Walton became an ordained minister and, like my father, an important citizen of Springfield, Ohio.

This was a time where there were few black players in professional football. It was very rare for black and white football teams to play against each other. Some my father's football heroes included Duke Slater and Jim Walker, who played for an integrated team at the University of Iowa, and Bill Willis, who played for the Ohio State University. It is quite possible that some of the later events in my father's life – some of the events caused by the advent of the trauma and the pain of World War II – sped up the process of integration. Many of my father's military experiences were ten, twenty or thirty years ahead of the rest of the country when it came to integration.

In the NFL, many of the football teams integrated far long before the rest of the country because like the military, sports teams are all about winning. Cutting-edge coaches were a little ahead of their time and would do whatever it takes to win. That is exactly the type of man that my father was – sure he stepped onto that football field and they called him "Li'l Abner." It is easy to think that he was called that because he was so country, so unsophisticated, and a simple man, but do not forget Lil' Abner's other great qualification: his most specific and memorable trait was his unbelievable strength.

My father's unbelievable strength more than anything else carried him through his military career and the rest of his life. He had both the strength of character and strength of body many other men would do anything to possess.

Dad as a freshman football player at Wilberforce University 1940

Wilberforce Football Team 1940

Wilberforce Football Team 1941

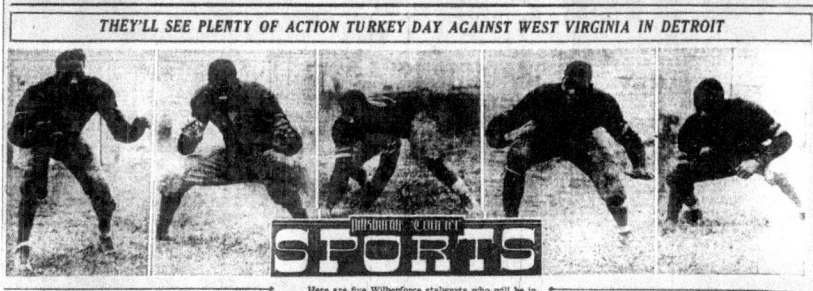

Wilberforce University's top linemen, as featured in the "Pittsburgh Courier," Thanksgiving 1941

6

FORT SILL

In 1941, the United States Army opened up a field artillery school at Fort Sill, Oklahoma. This was another response to the massive quantity of bodies needed to go to war on the Eastern, Western and Pacific fronts. There was a rapid need for leaders, and this is also one of the four locations where new men in the Army could go to basic training. We often forget what young boys experience between the call-up to the military and stepping up onto those boats in Normandy. Many of us only think of World War II as soldiers just being deployed, but there was a time in between where the United States had to rapidly expand their training.

My father's military training was unconventional, to say the least. As a college student in the ROTC program, he was on the path to becoming an officer from day one. We all know now that anyone who completes ROTC in the college requirements ends up as an officer rather than an enlisted soldier. At the time though, it was not sure for a black man; there were plenty of these men who did the same ROTC training with the same intensity, graduated from historically black colleges and were still only given high enlistment positions, but not made officers. It may have helped my father's belief system that his father-in-law to be, my Grandpa Warrick, earned a commission as an army lieutenant during the First World War.

World War II changed many things, and in a moment, I will share with you the story of the amazing unit my father joined, and how he showed the rest of the world what a black man can do when he puts his mind to it. Few monuments to the unit my father would later join still stand in recognition of the first time the American black man stood against the Nazis. My father would be there for the second time.

Fort Sill, where my father went to basic training, tells an interesting story of the US military. It was one the first places they opened to speed up the training, and as soon as World War II ended, they closed the training only to reopen it again for the Korean War.

My father was one of the last military trainees to see a horse pulling a cannon. When we think of horse-drawn artillery, we do not often think of them with cannons, and we certainly do not think this was still existing at the start of World War II, but it is something my father saw on the battlefield in 1942. It is a memory I always take with me.

In 1963, just twenty-one years later, they brought the horses back, but they only used them for a few local parades to muster up some of that patriotic spirit. Those horses no longer face the risk of going to combat.

ll-American Stars ationed At Fort Sill

T SILL, Okla, July 22— ... ting athletes from Negro have made the 31st Battaield Artillery Replacement Center. a ... ed proving for embryo artillerymen. ver has the battalion seen a collection of ... American ... as the ... up now on

ng the camp are Privates
... 612 East
street, Fort Worth, Texas,
College ... C. Wilson,
West Virginia of West
State ... and Milan
... 30th street,
... of Wilberforce, Xenia, Ohio.
... and quarter-
... the Negro
... last Fall
... year.
... second
... then
... in 1941.

... are
... 1st Battalion Shepard.
... third
Fall, and
Hervie
... tackle,
1941 and
in 1940,

... of the
... even last
... of captain the
... ety. The
... was chosen
... on list for
... in 1941, his
... elevated
... the 1942
... nine regular. Among
... 1803 Dalton, West
... and Samuel
... West Virginian men-
... was also
... ng cham-
... nia State,
... passing,
... and tackled Private
... elite in
... the
... third team
... at left half.
... ain of the

Hale Collects on Coast

LOS ANGELES, July 22— Fighting in the feature bout of a card featuring three main events, Billy Hale, Phoenix lightweight, won a close ten-round decision over Elroy Renteria, Mexico, at the Olympic auditorium here Tuesday night.

Don McLean, Chicago, dropped a decision to Enrique Bolanos, Mexico City featherweight, and Bobby Yeager, Los Angeles lightweight, scored an 8-round tko over Memo Llanes, Tia Juana, in the companion bouts.

1942 eleven. Qualls was also chosen on the Negro all-midwest first string in 1941 and 1942. Wilberforce team mates who are also training in the 31st Battalion are Samuel Crowell of Orlando, Oklahoma, right half, and Ross P. Barrett, 1322 Wheeling avenue, Cambridge, Ohio, tackle.

—VV—

Ex-Sparring Partner of Joe Louis at Army Air Base

ARMY AIR BASE, SALT LAKE CITY, Utah, July 22—"Joe Louis hits pretty hard and once he hits you, you might as well go home because you stay hit." Thus comments Pvt. Clinton Bridges of Detroit, Michigan, once a sparring partner of the champ and now stationed at the Army Air Base in Salt Lake City.

The 175-pound soldier started boxing in 1931 and won the Detroit Golden Gloves Tournament in 1932. After further successes in Detroit, he was sent to Chicago for the National and International Golden Gloves finals where he won the heavyweight crown. He started sparring for Louis in 1936 at his summer camp New Jersey and remained with the champ for three years but was forced to give up the active fight game when a huge dumbell rolled on his foot, breaking it. "I can still get around," he said, "but not like I used to."

Mentioned in the Pittsburgh Courier while at Fort Sill

7
WEST VIRGINIA

My father had an incredible, continually budding intelligence. I often think of him as an unstoppable object. Anything you put in front of him he would accomplish; he would overcome it or break directly through it.

He was invited to participate in a program that you probably have never heard of – it has faded into the annals of time because it only existed in the military for eighteen months. This program was called the "Army Special Training Program," and my father attended this as part of the pre-engineering school at West Virginia State College from 1943 to 1944. The purpose of this program was to prepare officers for the military.

At the time, the Army began to realize they did not have nearly enough soldiers to throw to the grinders of World War II. They took many young men and put them through an eighteen-month training program that once lasted four years. It was an accelerated training program and lest you think this was only for black – the ratio of white to black soldiers in this program was one hundred to one, unlike the rest of the military where it was eleven to one. The initial program was set with a maximum of one hundred and fifty-thousand soldiers, and one hundred and forty thousand of them were white.

Two hundred and twenty-seven universities around the United States participated in this program, training the new generation of officers that would be thrust into the war eighteen months later. Of these, only five are traditionally black colleges. In this program, the soldiers might be attending university, but they acted like soldiers – they wore their uniforms and they marched. When they stepped into the class, they stood at parade rest until their teachers allowed them to sit down. Many of those teachers were a little bit flustered, as there was very little warning that the program was coming to West Virginia. When these young, proud soldiers marched into the room in their uniforms, the teachers did not know the proper procedure to get them out of standing at attention. Eventually, they figured things out, and the learning was able to begin.

As it is often the case when the military and the civilian world meet, there was a little bit of friction and confusion. When you went to college and you chose how many hours of classes you attend, you would most likely choose between fifteen and eighteen per semester. Very few universes will let you spend more time than that in the classroom. My father could only look at my college schedule several decades later with a slight smile on his face. As a member of the Army Specialized Training Program (ASTP), he had fifty-nine hours a week of assignments, beginning with a minimum of twenty-five hours of class time and lab work combined with twenty-four hours of study, six hours of physical instruction, and five hours of military instruction.

Along with one hundred and forty-one thousand other men, my father wore a uniform and marched to and from class, unlike the other students on campus who wondered what our lives would hold. My father and the other members of ASTP knew that it was not a question of whether they were going to war, but where would they be sent; to the Pacific or would they be sent to Europe?

A fascinating detail about this program is that, just like anywhere else in the military, it was not voluntary. The only prerequisite was that you had completed high school and were between the age of eighteen and twenty-two. When you took the military tests, and they tested your intelligence using the military programs, if they determined you had the qualifications, off you went to officer training, and that was it.

West Virginia State College participated in this program from 1943 to 1944. They only had a single run of this program because, as with many things involved in the military and the government, politics intervened. My father joined the military at exactly the right time to go into an accelerated program to become an officer. As the military realized how many young boys fresh out of high school were being sent to war, they identified a need for more and more officers to lead those men in their small units. Though he would never say it, my father was lucky (or perhaps a better word is fortunate) to have been selected quite at the right moment.

Sometimes, opportunity comes where you least expect it. As I spent time growing up with my father on post after post, I began to understand that the military more than any other part of culture or government is pragmatic. The military's number one and truly sole focus is war winning. This is why the military desegregated far before anywhere else in the United States.

When my father graduated in 1943 and completed his training, he graduated into an Army that was still segregated; white students participated in ASTP in white colleges, and black students participated together in black colleges. Combined, there were one thousand four hundred and five black men who participated in ASTP. Of the millions who went to war fighting for our country, my father was elite – he was chosen out of one thousand four hundred and five. I am pretty impressed by my father for that one. Twenty years later, the Army had to explain why during the period right before they desegregated there were few black men in these programs and few black colleges selected – the ratio certainly did not match the ratio of the rest of the military which was about eleven white soldiers for each black soldier. There were a few assumptions, and part of them came down to segregated education.

At the time, many black men had limited educational opportunities, and not very many of them were able to learn to read, so their limited opportunities affected them over and over again. Part of this was some mistaken assumptions on behalf of the military about black colleges not being able to handle this program because they did not have qualified teachers. Later research proved this wrong – there are plenty of excellent black colleges beyond West Virginia that could have trained other promising young officers.

Far more fascinating is the information flow during this time. While the program was announced a year earlier, West Virginia State College only found out two weeks before my father showed up that they were participating in this program. As patriots, they were willing to adapt and scramble to prepare for this surprising influx.

It was not the first time this proud school participated in the military tradition. Founded in 1891, it was only a few years later that the school's principal would step down to go and fight in the Spanish-American war.

In fact, long before the advent of the GI Bill, where the military would pay for you to go to college and provide you the opportunity to become an officer or achieve your educational dreams, in 1899 West Virginia passed the so-called "Cadet Bill," which would provide up to sixty young men with free tuition and room, board, and books. Soldiers had always been a part of West Virginia State College, but with the ASTP they became a bit of a driving force, and the number went far past sixty and close to five hundred.

In July 1943, just two weeks after the president of the college received that warning phone call from the military, my father arrived on campus. I can only imagine the strangeness; my father, a Wilberforce student playing football like any other college student, called up by the military. He goes to basic training and then is sent back to college and told he has to complete his entire college degree in eighteen months instead of four years.

What a surprising and fascinating twist of fate. My father and one hundred and seventy-six other men were part of ASTP unit number 3537 on campus to take accelerated engineering courses. None of the soldiers knew they were part of an elite program; the ASTP had only been founded that year, and they were the first participating class (and as they would soon find out, they were the last participating class as well). They were young, they were thrown into a world they did not really understand, and they just went where the Army told them to go. My father simply took it as the next step on his path; he was stoic in accepting his fate and always looking for a way to turn into his best opportunity.

While the students were housed separately, the military boys were in a barracks style environment; they marched to and from each of the classes and to and from the dining hall. The education, the teachers, and the quality of experience were the same. While many of the military boys enjoyed spending time with beautiful young coeds, my father already had his heart set on my mother, whom he had met back at Wilberforce.

He had no time for this, no time for love, no time for frolicking – my father was preparing for war. He was training, learning, and studying to become a leader of men. When you are in charge of other men's lives, there is a great deal of pressure involved. It is both an honor and a source of terror – if you make a mistake, it affects more people, if you do things right, it saves more people. My father chose to throw himself fully into this program and give it one hundred percent of his focus, rather than falling to the distraction of young love. His heart had already set on the goal, and when my father set a goal in front of him, it never was a question of whether he would achieve it.

No one ever wondered if my father would graduate college, if my father would become an officer, if my father would come home from World War II; he was not that kind of man. It was only a question of when. When my father sent a letter to my mother asking her to wait for him to come home, our destiny became inevitable. My birth and that of my younger sister Rosalyn became inevitable. That was the force of my father's indomitable will.

The beginning of this program was rather fascinating. They were there to complete their basic engineering degrees and become officers with the ability to handle much of the new technology entering the military at the time. Tanks, submarines, airplanes, and bombers –

war always accelerates experimentation design, and man is brilliant at developing new more terrifying ways to destroy his own kind. The mix of students was rather wide; with such simple qualification, some of the students were eighteen, and some of them were twenty-two. My father had just left the college campus a few months earlier and was thrust back into the same world, so he was a little more prepared than some of the other students who simply graduated high school four years earlier and scored high enough on the military aptitude test.

Some of the students had to go through a remedial training (the training on how to sit in the classroom and go through an educational experience), but over time things smoothed out, as all soldiers caught the rhythm and accomplished what was required of them. Unfortunately, some of the men in this program could not quite handle the transition and went back to their previous units, but the majority of them held strong in. This experience was the same across ASTP – not every single one of those hundred and forty thousand white men completed the program either. When the military makes things tough, they make them really tough. The students who had already completed a year or two of college were more than prepared for this program, and they almost breezed through. Many of these students were the ones you had the time to date those beautiful coeds. As always, my father remained locked on his target.

The students were forbidden from playing college-related sports – which is probably the only way they could have kept my father off the football field. He looked at the fifty-nine hours a week of activities and said, "I still got a few more hours left to crush out on the old football field, step onto the old the gridiron, and show people what I am made of." Instead, only intramural sports were allowed. The soldiers were also allowed to enjoy a military ball; they would invite a young debutante to join them, and they would dress in their military uniforms looking their best, or as my father would say, "Looking spiffy."

In their hearts, these men always knew they were in a world apart; while they were enjoying a little bit of college, their brothers, their friends, and their former classmates were facing danger, getting hurt, and losing their lives for their country. As much as these men enjoyed being a part of this program, there was a part of them and a part of my father that felt guilty, that felt a little bit of, "Maybe I could make a difference over there right now," and I wonder whether this feeling planted the seed within my father that kept him in the military for twenty years after the end of World War II. This feeling that he could be the one to make a difference.

As it happens anywhere, when you put a group of young men together (and especially the military) they begin to develop a tradition. ASTP men developed a series of different poems and songs, and one of them, sung to the tune My Bonnie Lies Over the Ocean, went thus:

> *Some mothers have sons in the Army,*
> *some mothers have sons on the sea,*
> *take down your service flag mother,*
> *your son is in the ASTP.*

This song was a reminder that while they were in the military, they were not in danger, but eighteen months later they would be thrust right into the teeth of World War II.

During this time, they were filled with feelings of conflict. Of course, my father was glad not to be in danger – he could only think of going back to my mother and taking her hand in marriage – but at the same time, he felt guilty for all of his friends and fellow students at Wilberforce who were not invited into this program. Although my father did not express his emotions outwardly very much (he grew up in a world where this was frowned upon and looked down upon), he still had these emotions, and he still felt these things deeply. He just believed that most of your emotions are a private affair.

One of the benefits of this program was that the regular students and ASTP students were in competition with each other; this would drive them to play harder on intramural field and study harder in the classroom. This led to an educational peak in West Virginia State

College – everyone fighting for better grades revealed the surprising benefit of combining military and non-military college education. At the end of 1943, with the program only partially completed, the men knew they would be invading Europe in 1944, and with the coming invasion, they knew that the military would need a lot more boots on the ground.

While they went to class every day and they even jokingly called their program "All Safe Till Peace," they all knew that any day they could get the letter, the telegram, the phone call to say, "You have been called up, there is no time to complete this program. Pick up your gun, let's go." Although they lived with a little bit of guilt about their friends already in combat, they knew that it could be their turn at any moment. There was no guarantee that they would be allowed to complete the program.

On February 20th, 1944, they received a message that they called their "kiss of death" letter. In short, this letter said that in order to break the enemy's defense and assure the United States victory, study time was over (of course I am paraphrasing here, as you know the military is nothing if not eloquent within the letters). If you asked most men from this program, they would not remember the content of the letter, but they would remember the core message: you are going.

At the time, qualified numbers of ASTP could try and transfer into the Air Force, but shortly after receiving this letter, this program was shut down. Many of my father's classmates who made an attempt, knowing that the Air Force was far safer than the ground forces, were perhaps just wanting to fly in the sky like my father did. By the time they decided to change, it was too late in the Army close that door: "You are going to the infantry."

My father, along with the rest of his ASTP class, graduated on March 16th, 1944 to quite a rousing send-off. For all the student that had been in a competition for the past semesters, everything changed, because now the guy they were facing across an intramural field was going off to Europe to die and fight and protect everything you stand for and everything you believe in – to give you a chance that maybe they can finish that war before you graduate and before you have to pick up your rifle. There were quite a few tears shed on that graduation day, where my father graduated in far less than the eighteen months he had initially been promised.

Today, this program has been forgotten by the annals of history. Most people credit the atom bomb with a victory in World War II, not ASTP. Yet, had we not developed that bomb, we would have needed more soldiers, engineers, and doctors quickly accelerated and trained for this program, and it would have simply continued my father's love affair with education before he was sent off to Officer Candidate School.

8

OFFICER CANDIDATE SCHOOL

Officer Candidate School (OCS) is a twelve-week program designed to quickly and intensely prepare young soldiers. Mostly enlisted soldiers were being transitioned to officers; the outcome and freshly minted second lieutenants. This machine and program was mostly designed to convert mustangs. Mustangs are soldiers who come up to the enlisted ranks and then transition into being officers – unlike the traditional path, where officers either go to one of the military academies or they enter the military with a college degree often having completed the ROTC program.

We are all familiar with OCS as a program now, but it didn't always exist. It was first proposed in 1938, just a few years before my father joined the program. This program, like ASTP, was designed and developed as a response to what was happening in Europe, and what would shortly happen in the Pacific.

Although the Army is a large machine and a bit of a behemoth, sometimes it realizes that it needs agility. They developed a program specifically designed to quickly generate the officers they would need to grow the military and replace the staggering losses we faced during the Second World War. This was the only commissioning source, the only way of developing and building officers that could quickly adapt to emergencies. The other programs for developing officers all take four years. When war comes, you do not have four years to build and prepare your fleet of officers.

This program, which still runs to today, was specifically designed to add an element of agility to the United States Army. It was initially discussed and requested in 1938, though no action was taken until 1940, when a plan was finally developed, and only in 1941 did they actually begin running soldiers through the OCS program. As much as the Army wanted to develop an agile program for developing and building officers, it still took them three years from idea to implementation. It was into this program that my father was thrust when ASTP was shut down early. My father was always destined to be an officer.

Interestingly enough, each time he entered an officer training program, it was a little bit shorter. The first program was four years at Wilberforce. The second was eighteen months at ASTP, at West Virginia State College. Finally, he entered OCS at Fort Benning, Georgia, where they would prepare him to be an officer in just three months.

My father went to Officer Candidate School during one of the most fascinating and terrifying times in American history. Terrifying because at the end of it, when he completed his training, he would be sent straight to war to fight and possibly die for his country. Fascinating because he was at the cutting edge of military desegregation. At that time, the United States military enlisted soldiers were separated into black and white. Units were segregated and rarely fought together. There was simply no room or time for this with officer training.

Starting in 1942, shortly before my father was to enter his OCS class, the Army decided, "We are training them altogether; it is too important to put together and develop the best war fighters we can. We do not have time to develop two separate programs. It is too hard to maintain segregation and to win a war at the same time." Who would have thought that it would take a war like World War II to show that segregation was a waste of energy? If you asked my father about this time, he would rarely talk about the segregation elements; he was not really focused on other people and other people's experiences. He was only focused on his own destiny.

While for the first time he was training side-by-side with white soldiers, he did not see it that way. When he looked to the left and to the right, he did not see black or white; he just saw army green, and so did his fellow students at OCS school. There is nothing like the fear of the great tyrant to make those little things so insignificant. It is not such a big deal what color the guy next to you is when there is someone out there who wants to enslave both of you.

OCS was the Army's first formal experiment with integration. Black and white candidates shared quarters; your roommate was no longer determined by the color of their skin, but by the first letter of his last name instead. Unfortunately, while the training was integrated, the units were not yet. When my father completed his OCS training, finally becoming the officer that he knew was his destiny, as a freshly minted second lieutenant, he was to join the 371st segregated and black-only unit. He had a taste of integration, and it was quickly taken away.

AT HOME AND OVERSEAS
What the GIs Are Doing

WASHINGTON, D. C.—The Department of the Army announced the appointment of twenty-three Negroes as second lieutenants in the Regular Army.

The appointment of these officers raises to eighty-two the number of Negro officers in the Regular Army. The following officers were appointed: Ross P. Barrett, 1322 Wheeling Avenue, Infantry, Cambridge, Ohio; Robert L. Barringer, 1050 N. Fifth Street, Infantry, Birmingham, Ala.; Jew D. Boney, P. O. Box 777, Rockdale, Tex.; William P. Bullard, 121 Penn Street, Infantry, Martins Ferry, Ohio; Samuel F. Sampson, P. O. Box 7102, Houston, Tex.; George M. Shuffer Jr., 1551 Truman Street, Infantry, San Pablo, Calif.

James A. Bailey, 676 Bowman Avenue, Field Artillery, Columbus, Ohio; Spencer M. Bracey, 733 Walnut Street, Infantry, Camden, S. C.; Orlando V. Brown, P. O. Box 263, Field Artillery, Tridelphia, W. Va.; Preston A. Davis, 1233 Marshall Avenue, Field Artillery, Norfolk, Va.; Paul J. Dickerson, 612 Booker Street, Coast Artillery, Charlottesville, Va.; Oscar J. Harrison, 821 Chapel Street, Transportation, Norfolk, Va.; William J. Nelson, Route 1, Box 13, Infantry, Columbia, S. C.; George E. Peters Jr., 903 Summit Street, Field Artillery, Wheeling, W. Va.; William B. Proctor, 307 N. McComb Street, Transportation, El Reno, Okla.; Charles E. Scott, 103 Emilin Terrace, Field Artillery, Lawnside, N. J.; William R. Smithea Jr., 2109 Bainbridge Street, Transportation, Richmond, Va.; Rufus C. Streator, Route 2, Box 767, Infantry, Hartsville, S. C.; Robert L. Turman, 1136 Tuscalousa Avenue, Infantry, Gadsden, Ala.; Albert L. Walker, Route 1, Box 246, Infantry, Crockett, Tex.; Lonnie R. Williams, 400 E. Miller Street, Transportation, Gurdon, Ark.; Richard A. Williams, Route 2, Box 5, Infantry, Elloree, S. C., and George Wright, 808 Bladen Street, Infantry, Beaufort, S. C.

* * *

KITZINGEN, Germany — With more than 500 men assigned, Headquarters and Headquarters Company of the 7871st Training and Education Group, is one of the largest administrative units in the European Command.

* * *

The company is commanded by Capt. Merle J. Smith, Greenville, S. C. A graduate of West Virginia State College, Captain Smith received his commission as a second lieutenant in July, 1944, and received his captaincy in 1946.

Captain Smith will return to the United States soon for reassignment. His wife, Mrs. Jacqueline Smith, and their son, Merle Jr., are presently living in Kitzingen.

First Lieut. Richard H. Nance of Columbia, S. C., is the executive officer of Headquarters Company. He received his commission in June, 1943, and was promoted to first lieutenant the following year. During the war, Lieutenant Nance served with the Ninety-second Infantry Division at Fort Huachuca, Ariz., and in the Mediterranean Theatre of Operations. He was graduated from South Carolina A. and M. College in 1943.

First Lieut. Duncan Williams of Greensboro, N. C., is the motor officer for the 7871st Training and Education Group. He entered the Army in 1942 and served with the Ninety-third Infantry Division in the South Pacific during the war. Before entering the service, Lieutenant Williams attended Agriculture and Technical College in Greensboro, N. C.

M/Sgt. Horace A. McCray, of Philadelphia, Pa., first sergeant of Headquarters and Headquarters Company, served with the 370th Infantry Battalion at Grafenwohr, Germany, before receiving his present assignment.

* * *

FORT ORD, Calif.—Corp. Mary E. Penn, of Washington, D. C., a member of the Wac Detachment, 6093rd Army Service Unit here, was recently designated as an honor student at the graduation exercises of the Food Service School sponsored by the Fourth Division Artillery at Fort Ord. She is the first member of the detachment to receive an honor award at the school.

Corporal Penn joined the Women's Army Corps in April, 1945, in Washington, D. C., and received basic training at Fort Des Moines, Iowa. Her permanent home is at 515 Q Street, N. W., Washington, D. C.

* * *

SECKENHEIM, Germany—Corp. King Baskin of St. Louis, Mo., was recently selected "Soldier of the Week" during a retreat parade held at Hammonds Barracks in Seckenheim.

Corporal Baskin's parents, Mr. and Mrs. Vernon Baskin, live at 3835 Tate Street, St. Louis, Mo.

* * *

KITZENGEN, Germany — Col. Henry C. Newton, commanding of-

The 82nd Black Officer in the United States Army

9

WORLD WAR II

My father left Officer Candidate School a freshly minted second lieutenant and was assigned to the 371st Infantry Regiment of the 92nd Infantry Division. At the time, he was one of only eighty-two black officers in the entire United States Army. One of his colleagues was Edward Brooke from Massachusetts, who was destined to become a United States Senator. He left behind his integrated officer training to take a position as an officer in an all-black segregated unit. My father fought with this unit from 1944 to 1945, earning three battle stars and beginning a collection of medals that would last another twenty-three years. He spent a full twenty months in Italy and, like many soldiers from that time, he did not talk about it much.

The 371st was a unique unit, and thus it faced some unique challenges – there were only three all-black regiments in the entire military. Because my father was a little hesitant to share his stories about his military career – especially the combat he saw and the terrors he faced – I have learned much of my father's history from his contemporaries, from his younger sister, and from online research. We know what happened because the military keeps extensive and detailed records, but the journey of the unit is not always the journey of the man.

When the 371st regiment was founded, the United States was in the throes of another world war. At the time, they never expected to fight World War I and World War II only twenty years apart. The 371st was one of the first times the United States fielded a large number of black soldiers; at the time, they were part of the 92nd Infantry Division, colored. That's right; they specified every time it was a black division. They came to France at a time when France needed a great deal of help, and they were immediately transferred to French control during World War I. This is why the unit has many medals and decorations that are French rather than American. Although the United States was willing to field black soldiers and build a black division, they were not quite ready to lead them.

During World War I, they were extremely well-decorated and achieved many victories when fighting for the French. They received many unit and individual citations and awards including the Croix de Guerre and the Légion D' Honneur. Additionally, they were awarded the Distinguished Service Medal from the United States. Corporal Freddie Stowers was the only black soldier in World War I to be awarded the Medal of Honor. Despite all its disadvantages, this unit achieved some amazing things.

The French were so proud of the unit that they established a monument that still stands to this day, despite the attempts of the German artillery to destroy it. While they damaged it, and the damage remains, the names of the black soldiers who fell during World War II to fight for freedom both in America and in France still stand strong and not forgotten.

When my father joined the unit in 1944, it was not an easy time to be a black man. Many men in the unit were suffering from a lack of education and a lack of training, and the unit was at a major disadvantage compared to other units in the military. Even though the United States military was beginning to desegregate, education would remain segregated for quite a while longer. That meant that soldiers who came from inferior or disadvantaged education would continue to suffer, even though the Army would offer them equal opportunities. It was a challenging time, and my father fought to rebuild, strengthen, and re-educate this unit, to catch them up to the readiness of the white units in their upbringing and education.

The 92nd Infantry Division was reactivated for World War II on October 15th, 1942. At the time, there were only three African-American divisions, but only the 92nd would serve as a full combat division. I would love to tell you that World War II was a breeze for this division and that those colored soldiers showed everyone else in the Army just what they were capable of. Unfortunately, they had a bit of a checkered career with some wins and some losses along the way, like many other units. They had a lot to overcome, but they did one thing that changed the course of American history and altered our military forever.

During World War II, they were the only division to become fully integrated. While this division still exists, it is no longer black only. They set the standard, and they are the reason why in the military, the color of your skin does not matter anymore; in the Army, everyone bleeds green. From 1942 to 1944, shortly before my father joined the unit, they went through a great deal of remedial training, fighting to bring the soldiers up to the levels of literacy, training, and readiness that they would need to enter combat on even footing with the rest of the soldiers in the US military.

There were many problems with training this unit. Because of strong feelings of racial disharmony, the unit was unable to train together for a very long time – they had to train separately at different bases. No other unit suffered from this challenge. For eight months, the training was conducted separately, but how can you fight side-by-side with someone you have not trained with? How can you prepare for maneuvers that you have never practiced together? After eight months, in late 1942, the unit transferred to Arizona, where they could finally train together. When they were training in the desert, this unit, which was already using a buffalo patch on the uniform, adopted a live buffalo as a mascot; that's how they came to be called "buffalo soldiers." Perhaps you have heard that term, and now you know my father was part of that unit.

It was a time of conflict within the unit because many of them got the lowest possible score on the military intelligence test, and more than a few of these young boys were illiterate. They were the product of the separate but equal educational system that somehow never provided equal results. It is hard to inspire soldiers to fight for a country that does not treat them like people.

My father was of a different mind. He felt that he could use the military to gain every advantage possible that the rest of society did not offer. He knew that the military would desegregate before everyone else. As much as the union military was going to use him and his fellow soldiers to fight for their ideals, he was willing to use the military to achieve his own personal objectives – to demonstrate not through words but through action that he was just as good as anyone else, by being a brilliant commander and field tactician. When bullets are flying and bombs are crashing down around you, no one cares what color you are; they only care if you can keep them alive for a few minutes longer, if you can lead them to victory, if you have what it takes to achieve that next military objective.

In the fall of 1944, the division deployed to Italy with my father. They were assigned to the Fifth Army following the fall of Rome, and they were to push onwards to the Alps – those same great mountains that Hannibal once crossed in the other direction to take Rome. My father and his men were sent to go back in the other direction, to reverse Hannibal's journey. Although I imagined it much as a child, unfortunately, my father's unit included no elephants. This was the first time my father faced a challenge that was a little bit beyond him. He was only a second lieutenant, and he had been through several partial training programs. He had not completed his college education nor ASTP, and he was bootstrapped to Army Community Service into the officer corps. While the 92nd division was fighting on the Italian coast as part of the Fifth Army, their unit suffered from more struggles than any of the other units. There were some soldiers that stood out above the rest, but lack of leadership and lack of training led to a lack of military discipline. It is hard to point the finger at any one person, and it is the result of report training regime.

Perhaps if they kept the unit alive and running from World War I to World War II, they would have had higher officers with the necessary experience. Part of me thinks that these initial troubles are the reason the military decided to desegregate all units. Perhaps a little suffering and a little negativity were necessary on the path to achieve something much greater. The Army realized they needed to mix these soldiers together with everyone else and start treating all soldiers equally. That first trial with mixing the training at Officer Candidate School showed that it could be done – it demonstrated that when you put a significant goal in front of people, their little disagreements disappear and fade away.

We have all seen loads of those movies about high school football desegregation, where the white and black players start agreeing with each other. Those took place ten or twenty years after my father's experiences in the military. I hate to get bogged down in the worst part my father's history, but sometimes the darkness is necessary to understand the significance of the light. After several setbacks, the 92nd division was assigned to cross the Cinqual Canal, in February of 1945. Despite having a great deal of support from artillery and the Army Air Force, the 92nd got stuck.

For three days, they could not cross that canal, and eventually they retreated. Out of that loss came something amazing; the soldiers who were still too far behind and needed more remedial training were separated from the rest of the unit – they were sent back to America for more training. To replace them, the War Department assigned two new regiments; one with the 473rd and the other was the 442nd. The former was a white unit, and the latter an all-Japanese-American unit that eventually became the most decorated unit in World War II. The 92nd went from all black to equal parts black, white, and Japanese American. At a time when Japanese Americans back home were interred and suffering their own challenges, many in America wondered if they were turncoats, traitors, or spies – no matter how many years they had been living amongst each other.

These three regiments worked together to show the world and the military just what people are capable of. They showed that no matter how badly some people are treated at home, they will fight for their country. This new regiment became quite successful and began to push back and take more territory; capturing the cities of Massa, La Spezia, and Genoa by the end of the war. In 1945, when those remedial soldiers back in America were ready to come back to war, they returned to the unit, and the white and the Japanese-American regiments went back to their previous divisions.

My father participated in not one but two great experiments in unification and desegregation. While it took the military a while to fully desegregate unit to unit and eventually make it every single soldier, they would never again field a segregated division. In 1948, when reviewing the entire United States experience of World War II, President Truman looked at our wins and losses, our successes and areas for opportunities and growth, and he told the military, "Enough, desegregate now."

At the end of World War II, the 92nd division was stood down and never returned; the United States no longer fields all-black divisions. In 1950, with the growing needs of World War II, the military finally completed their desegregation process and asked my father to pick up a rifle and sent him to Korea.

Lt. Ross P. Barrett, Fort Dix, NJ, circa 1949

10

SEMI-PROFESSIONAL FOOTBALL

My father was a busy man, but he was not one to give up on his dreams. One of my fondest memories was that he went back to his love of football in between World War II and the Korean War. In the late 40s/early 50s, he played semi-professional football at Falmouth, Massachusetts. While stationed in Camp Edwards in Massachusetts, he found time to rekindle his love for the old gridiron, but because he was not the young man he used to be, his opportunity to play in the NFL had passed by.

The NFL was now fully integrated, and the rest of football was following suit, so more opportunities were headed his way. Although he was no longer a lineman, he was still strong. There was a clipping he showed me from the Falmouth Newspaper of him kicking a long field goal to win a football game against somebody. I have a vivid memory of listening to the story about my father kicking the game-winning field goal. I remember when I did some military training at the Air Force ROTC summer camp, at Otis Air Force Base up in that Cape Cod area, about twenty years later. I was up there during the summer of 1970, and I met some people who still remembered that very field goal. The pride I felt in my heart in my heart is something that still sticks with me all these years later.

Just because war, marriage, and kids enter your life, it does not mean that you have to give up on your passions. It doesn't mean that you have to give up doing what you love. Even if you are not as strong as you used to be and you cannot play the position you used to play, you can still find a way to enjoy sport. My father used football as a way to leverage himself and improve his life by getting into a good university, but he also loved the sport so much that he kept it as part of his life, even as he prepared to go off to war yet again.

11

MILITARY CAREER

My father decided in 1945 to serve out an entire twenty years in the military. There was massive demilitarization going on, and anyone who wanted out could get out after the victory in World War II. Many of his friends and comrades in arms made that decision and went back to their previous lives. My father was a man who was always looking at the long game, and he saw the military as an opportunity and a lever to get access to levels of education he would not have been otherwise able to achieve.

The rest of the world and the rest of the education in America would not desegregate for another twenty to twenty-five years, and my father did not want to wait that long. He knew that inside of the military he would have access to undergraduate and postgraduate opportunities – something that non-soldiers (and especially black non-soldiers) would not have access to for a very long time.

While my father loved his country and certainly enjoyed serving his country as he fought in two foreign engagements, he never let go of his sense of self. He always saw his relationship with the military as a two-way street, "I will do these things for you, and you will do these things for me." This cooperative mindset it is what caused him to not just endure but to thrive in the military – to achieve great things and have amazing experiences.

I do know that my father had one regret, and that was he was not in the right place and time to join the Tuskegee Airmen. He always had a passion for flying, and if he had made that transfer in between ASTP and OCS training from the Army to the Air Force, he could have had a real opportunity to sit on a plane and fight for his country from the sky. He was not a man to give up. He was simply a man to wait with a great deal of patience; he knew that one day he would get to fly for his country and he was more right than we could have ever predicted.

As early as the mid-1950s, he began to talk with the civil air patrol, and he discovered a branch of the US government that was parallel to the United States military but not really part of it, where civilians would fly planes on behalf half of the local and federal government. Sometimes they would fly these planes if a child was lost in the mountains, or to keep track of a wildfire. They also flew out over the coast during World War II to keep their eyes peeled for submarines and U-boats – a task that could not have been more critical, as after the surprise attack on Pearl Harbor, many Americans lived in great fear of a Japanese submarine surfacing off the coast California or Washington. The civil air patrol filled in a gap and flew over those waters in the hopes of being the first to notice and respond to this great danger. Fortunately, this fear never came true, but these brave and often forgotten heroes were people that my father admired. His fascination with the civil air patrol percolated in the back of his mind for a very long time, and it is something that would come to fruition nearly twenty years later.

He did have a few experiences flying with the Army, and I have seen some old photographs and slides of him standing next to US Army Piper Cub. It was never enough for him; if he had been born a little bit later, he could have accomplished so much, and I have no doubt that if he had been born twenty years later, he would have been a fighter pilot flying over Vietnam. He could have been one of the Tuskegee Airmen, but sometimes history conspires against us, and we find ourselves in between two great opportunities.

In those moments, it is tempting to it consider quitting or saying that fate must be against us. My father's life and his story are a testament against that; instead, he was proof that you could fight against history, and even when the opportunities were not there, you could create and make your own. My father found his own way into a cockpit just a few years later.

Dad - somehwere in Korea - circa 1952

The Barrett Family - 3113 Gwynns Falls Parkway - Baltimore Maryland, circa 1956

Barrett Children - Baltimore MD - circa 1956

Dad meets us at the Frankfurt Germany airport - Dec 1957

12

MARRIAGE

Before going to a second war, my father took a little time for himself. If you remember, he asked a certain special lady whose father was Professor Warrick from Wilberforce University to wait for him. As she was one of the most beautiful girls on campus and quite eligible, many men asked for her hand, but she waited for the hero to come home. On June 23rd, 1946, my father Ross Paige Barrett married Allyson Lightfoot Warrick.

The beautiful young bride and the second lieutenant (who wasn't quite so freshly minted anymore) were married at her family home. Now that the war was over, the real benefits of being a man in the military were about to bear fruit for my father.

Grandpa Warrick, who wasn't particularly impressed with Dad in the beginning, got to a point where he was finally alright with his daughter's choice. My father was a simple guy, not a city slicker, and I guess grandpa felt comfortable in trusting his daughter to him.

I was the first child. I was born in 1949. They bounced around couple military assignments between 1946 and 1949 – Fort Benning in Georgia and Fort Bragg in North Carolina. At the time I was born, Dad was stationed at Fort Dix, New Jersey, where he was an infantry instructor. While they were at Fort Dix, Mom did some teaching at a place called the Bordentown Institute, which wasn't too far from Fort Dix.

Soon after I was born, I remember there being an assignment to Camp Edwards in Massachusetts, down in Cape Cod. During the time they were there, Mom did some teaching, and Dad was doing some due military infantry stuff. He also played for a semi-professional football team out of Falmouth, Massachusetts.

Then the Korean War started in June 1950. In 1952, my father was assigned to the 24th Infantry Regiment of the 8th Army's 25th Infantry Division. He was sent to fight in the Korean Theater – he had fought in Europe, and now it was time to fight in Asia. Again, he participated in three major engagements and earned three battle stars. During World War II, this unit was moving so fast across the Pacific Theater, hopping from island to island on its way to press into Japan, that it earned the title "Tropic Lightning." From a Buffalo soldier to a member of the Tropic Lightning, my father always seemed to find his way into storied units.

Before my father joined the unit, they had already shipped out to Korea. Their first assignment was to protect the port city of Pusan from North Korean encroachment. They were to hold this position so that the United States could land support troops. The last thing we wanted to do was losing ground to the North Koreans. Sea-based invasions are hard, both in terms of strategy, material, and life. The Tropic Lightning lived up to their reputation and held that port – they held back the North Koreans and received their first Republic of Korea Presidential Unit Citation. This was far from the end of their success in the Korean conflict; it was only the beginning.

On January 15th, 1951, Tropic Lightning was ordered to push north and begin retaking territory. Pushing the North Koreans back by February 10th, they captured Inchon and the Kimpo airbase. This was the first of several assaults on the opposing Chinese and North Korean forces.

My father's division was then assigned to participate in Operation Ripper, where they pushed the North Koreans and the Chinese back again to the other side of the Han River. We all know how the Korean

War ended – we pushed them back to the 38th parallel – but knowing how a war ended is not the same thing as being there and certainly is it not the same as being a troop on the ground when the bullets are flying.

The unit then participated in Operation Dauntless, Detonate and Piledriver that spring, which secured the Iron Triangle and put the United Nations in a strong position to negotiate and prepare for the end of hostilities. During this time, the thirteen men in this division would earn that highest-level claim that is the Medal of Honor.

My father came home from Korea with the sole goal of continuing his career in the military, growing his family, and hopefully avoiding any more conflicts. I grew up going from assignment to assignment. My Dad was certainly very busy as a military person, and I didn't get to know him that well until after he came back from Korea. One thing we used to do a lot that was my first introduction to professional football. Dad used to enjoy taking me to see the Baltimore Colts play at Memorial Stadium. I recall that some of my favorite Colts were Lenny Moore, Alan "The Horse" Ameche, Gino Marchetti, Johnny Unitas, Raymond Barry, and Gene "Big Daddy" Lipscomb.

My father had drawn blood and shed blood for his country, earning a Purple Heart amongst many other commendations. He was now ready to participate in the other part of the military. He had joined ASTP as part of a pre-engineering degree – he had a passion for the industrial arts in high school, which unfortunately have disappeared from many classrooms these days. They would become a larger and larger part of his life as he continued his military career and transition onwards.

Grandpa Warrick - The Father in Law

Announce Engagement Of Allyson L. Warrick

WILBERFORCE, Ohio—Prof. and Mrs. Ennis Champ Warrick of Wilberforce University, are announcing the engagement and approaching marriage of their daughter, Allyson L. Warrick, to First Lieut. Ross Paige Barrett.

Miss Warrick is a local public school teacher, is a graduate of Wilberforce University and is doing advanced study at the University of Michigan. She is a member of the Delta Sigma Theta Sorority, the professional sorority of Phi Delta Kappa, Sen Mer Rekh Honorary Society, and Zeta Sigma Pi, national honorary social science society, and is the newly-elected director of the Michigan branch of the American Teachers' Association.

Lieutenant Barrett is the son of the late Lee Barrett and Mrs. Nel-

MISS ALLYSON L. WARRICK
... *Pretty Bride-elect*

lie Barrett of Cambridge, Ohio, having pursued his college work at Wilberforce University where he starred on the fame 'Force football eleven and in inter-fraternal sports. He is a member of the Omega Psi Phi Fraternity.

He recently returned to the States on a Temporary Duty Leave of forty-five days. Formerly with the 371st Infantry Regiment, he is now with the 110th Quartermaster Battalion. Lieut. Barrett has served overseas in Italy for twenty months. He wears the Purple Heart, the Combat Infantryman's Badge, the Good Conduct Medal, and the Mediterraneal Theater of Operations Ribbon with two battle participation stars.

The wedding will take place on June 23 at Wilberforce University with Bishop Reverdy C. Ransom officiating.

—V—

Engagement Announcement in the Pittsburgh Courier

Wedding - 6-23-1946

13

FINAL MILITARY

By the time my father was finished with his extensive military career, he had received the first and second common infantryman's badge, an Army Commendation Medal, the Bronze Star, and the Purple Heart. He retired from the military on June 1st, 1963 as a captain – a rank that, when he was born, most in America would have considered impossible for black men to attain. At such a young age, he had already achieved the impossible over and over again.

One of my father's other passions besides flying was passing on his knowledge to the next generation. Education and sports had lifted my father up from humble and limited beginnings, and he believed education could lift the next generation even farther. I have certainly risen very high on the lever of his shoulders, indeed thanks to the effort my father made and the challenges he faced. His belief in education was right, and he saw it as an opportunity.

In the military, he first expressed his passion for education and teaching by becoming an assistant professor of military science and tactics at Morgan State College in Baltimore, Maryland. As he went through other assignments in the military, sometimes he was in charge of morale, but often he would circle back to passing on and guiding the younger soldiers – helping them keep their careers on the right path and see the opportunities in front of them. One of his Army ROTC students at Morgan State was Earl Graves Jr., who was destined to found and publish the renowned "Black Enterprise" magazine.

After three years in Pirmasens, Germany, from 1957 to 1960, he finished his career as a chief of the operations and training branch from 1960 to 1963 at Fort Benjamin Harrison, Indianapolis. One of his most exciting adventures during this time was meeting many of the Indianapolis 500 drivers when they came there for their annual competition and participated in local safety training courses. I recall that my father specifically spent some time with Lee Wallard, the winner of the 1951 Indianapolis 500.

Sometimes, little stories and things we do can get lost in the annals of history. My father wrote a report a long time ago that was one such thing. This has recently resurfaced and been quoted in multiple books; it was an extensive paper called: "The Role of the Ground Liaison Officer with a Tactical Control Group and Close Air Support." That is a very long title for an unbelievably critical position. Much of what he covers in that book appears in many modern military non-fiction works.

This position, in the simplest terms, is about the soldier calls in support for the planes. It is very important that the soldier on the ground and the pilot in the sky are speaking the same language. If you have spent any time around pilots and soldiers, you know they have a very different view of the world, they speak very different slang, and they see the world very differently. The pilot sees the world from the bird's-eye view, while the infantrymen (the grounds man) sees it from the ant's-eye view. They can only see a very small distance around them, so when giving grid coordinates or explaining their position, which is critical in requesting air support, there are moments where these messages can get lost in translation, and my father's work has become one of the seminal books on this very critical topic.

My father was a wonderful man who saw the military as an

opportunity to increase his education but also a chance to meet interactive people he would not have otherwise had an opportunity to talk to. He well enjoyed meeting those Indianapolis 500 race car drivers; he also got to spend some time with Buddy Baer, brother of the boxer Max Baer, who had fought Joe Lewis. Buddy Baer was a boxer himself, and he was a star of an old movie called *Quo Vadis*. In that movie, you can watch Buddy Baer wrestle a bull. When he was in Italy, following the hostilities of World War II, he developed a love for opera and classical music.

Working as part of the morale and vision leading team for a base means handling celebrities and USO tours – all of those aspects of entertainment that the military provides for the soldiers to keep them entertained and on board before they have to step back into combat. Morale is a critical factor, as units with low morale perform on the battlefield far below units with high morale.

As a young child, my father was, of course, my hero, and it was always fascinating to see him interact with different celebrities. He was a bit of an unsung hero, but his units and the soldiers looked up to him. The men he led were proud to have him in their lives. Sometimes, as a child, we just want to meet that race car driver, that boxer, or that movie star.

Though he was not a wealthy man, my father found a way to bring some amazing opportunities into my life and inspire me to achieve my own success.

Morgan State

14

EDUCATION PART II

My father's educational path sometimes reminds me of a puzzle. If the traditional path of education was designed on a puzzle, and then someone threw it against the wall and smashed the puzzle to a thousand pieces, that would be my father's education: completely disorganized. The end result is the same, but the structure is quite surprising. Over and over again, my father would go back in an effort to try and complete his college education. Because of the unique time in which he joined the military, he was a highly ranked officer who never actually graduated from a university, although he had started over several times, first at Wilberforce and then at West Virginia State College with that accelerated ASTP.

He continued to grab educational classes as often as he could; he studied at the University of Maryland, the University of Tennessee, Ohio State University, and Kent University, until he eventually graduated from Central State University, which was originally part of Wilberforce. Despite attending classes in so many different places, his life came to a full circle, and he completed his education exactly where he started it.

With this bachelor's degree from Central State University, my father had finally achieved the first step in his life plan just a few years or maybe a few decades too late, but he wasn't done yet. He immediately went on to complete his postgraduate work and a Master's degree at Kent State University. In 1965, with a freshly printed bachelor's degree in his hand, he began to teach structure and industrial arts at Springfield South high school. If you are little bit younger than me, perhaps you don't remember these programs, or maybe you remember a program called "metal shop" or "woodworking." These are courses where young students learn to work with their hands to begin to experience creating and working with metal, plastic, and wood. Unfortunately, these days we are surrounded by objects or machinery, but we have no idea how they work; the average person has no idea what is inside their television, and they certainly do not know what is under the hood of that car anymore. My father grew up in an educational system that taught you how to repair the things in your house; how to understand what is happening in the electrical wires that run between the toaster and your wall. He had a passion for the industrial arts – he was in the industrial arts club way back in high school.

My father would complete anything he started, and so he went from a member of his high school industrial arts club to teaching it just thirty years later. His teaching career was short-lived because he was so beloved by his students and he had so much respect (as you would expect from a captain in the military) that he was asked to enter the level of administration. When you have a man like my father on the team, a man who has led soldiers for so long, you cannot leave the classroom; you need him to lead – he was born to lead, and every task he entered, he eventually led.

15

HIGH SCHOOL LEADERSHIP

My father was always pulled to the top, almost like there was something magnetic about his leadership. No matter what company he was in, he was always drawn to leadership. In teaching, he was someone who had faith in passing on what he knew to the next generation, and other people always saw it within him. It is very hard to lead men in not one but two wars without becoming a natural leader, and my father exemplified this.

He turned to his passions over and over again. During the summer, he was always dedicated to teaching, and he was an instructor in mathematics at the apprentice training Department of International Harvester for seven years. He also worked in plant protection. They built many trucks at the Springfield facilities. Their mechanics and engineers in the mathematics departments were very important because when you are selling a truck that is worth a hundred thousand dollars or a million dollars, the last thing you want to have is a mistake or flaw in the design or engineering process.

My father's passion for industrial arts led him to this company, and he continued to grow there for many years during those summers. In 1964, he was an industrial arts professor in high school. Within four years, in 1968, he would become the assistant principal. If you know anything about educational systems, the jump into management usually takes far longer than four years. My father was not bucking for the promotion – he just had a passion for teaching industrial arts, but his leadership was needed.

When the call came, he stepped up. My father now has a plaque on the wall; they remember him at the school because he made such a difference. Even though he has not taught there for decades and decades, he is still remembered. He was the first black assistant principal at South High School. Even after he retired many years later, in 1982, he continued to volunteer and help.

When my father was not teaching or being the assistant principal, he loved to help with sports and athletics and serving as an official, particularly in football and basketball. He was incredibly driven, he always accepted the next challenge in front of him, and he led through example. These days, so many of us look at retirement at sixty-five, seventy-five or eighty, and we think that is the end of our story; we think that we can have one career and when it is finished, we are finished. My father is a demonstration that one man can complete career after career, and never quite finish the journey. He always found that next challenge around the bend.

His old school continued for many years to honor his name by offering a scholarship every single year to the student who most personified his ideals. What a way to be remembered.

One of Dad's favorite pictures of himself as an assistant principal at Springfield South High School circa 1975

Dad with some fellow South High Faculty members, circa 1965

16

CIVIL AIR PATROL

While all of these other things were going on, my father still found time to resume his flying, maintain his qualifications, and join the civil air patrol. He found his own sideways path into the United States Air Force, although he was not officially part of the Air Force – you can think of them as the civilian counterpart to the United States Air Force. In World War II, when the British soldiers failed at the first attempt to invade France at Dunkirk, thousands and thousands of local fishermen and ferry men and anyone who owned a boat came to get their soldiers home.

In the civil air patrol, it is very similar. They first began as the eyes in the sky for those German or Japanese submarines during World War II. Then they grew and started taking care of people in different ways. If someone were lost up a mountain, the civil repair air patrol would get the call, and it would fly to find them.

My father was responsible for more than a few critical rescues as part of the civil air patrol. He made such a difference that Springfield Cadet Squadron 702 of the Ohio Civil Air Patrol is now named after my father. He did not choose a name himself; it was after his death that they wanted to memorialize and remember him forever. Everywhere my father set foot, someone put up a plaque with his name on it. He could not touch the world without making a difference.

The civil air patrol is a force partner and auxiliary of the US Air Force. They are there to search and find the lost, provide comfort in times of disaster, and work to keep the homeland safe. Even when he stopped being a soldier, my father became a pilot. Even when the Air Force and the opportunity to transfer from ASTP in the Air Force was shut down, he found another way to fly for his country.

In order to be a member of the civil air patrol, you need to have a plane to fly. My father was not an extremely wealthy man, so he did not own a plane at first. He leased a plane in exchange for his work, ability, and skills – his industrial arts ability enabled him to maintain a plane. He worked his way up through the civil air patrol and eventually owned his own plane.

My mother eventually sold it on after his passing, but that plane is still out there taking flight. It was a Grumman Tiger; you can currently find it up in New England, where its present owner lives, just no longer in service as part of the civil air patrol. My father took such good care of his plane that thirty years after his death, it is still up there flying. I think that plane is still keeping an eye on us from up there.

Dad with his Gruman Tiger 388

Dad with members of the Civil Air Patrol

17

ORGANIZATIONS

When you look at all the things my father accomplished and all of the activities in which he participated, it is hard to believe that there is still more I can add to this story. It is hard to believe that one man could do any more the things than what I have already listed, and yet he found a way to do it.

He was a member of the Retired Officers Association, Phi Delta Kappa Education Fraternity, and the Omega Psi Phi Fraternity. He was also a member of the Dayton Opera Association, the Springfield Symphony Orchestra, and the Metropolitan Guild. He served on the board of United Way, the City Hospital, the Museum board, and Center Street YMCA, NAACP, and the Urban League. On top of all of that, he was licensed as a sports official in Ohio and Indiana for both basketball and football, and in Indiana, he also officiated baseball. He refereed hundreds of games and made hundreds of friends along the way. I have also seen pictures of him as a Prince Hall Mason, although I do not recall him participating in any Masonic activity during my lifetime.

The story of my father and the message of this book is one of possibility. In life, we often look at our limitations, we participate in a few hobbies or play few sports in high school. We walk into our careers, and we stop doing other things – we stop making new friends and having adventures. How many of us would consider starting a new career at forty, fifty, sixty or seventy? My father was an unstoppable force, and he seems like one of those people that simply never got tired.

When you are going through those moments in your life where you think about what you can do and what you can accomplish, I want you to remember this book not only as the testament of one son about the greatness of his father, but most of all as an example of how one man can make a real difference. One man can bend the universe and affect the world. Whether you are a man or woman, you have the ability to do great things. There are more adventures waiting for you around the corner; your story it is not finished, and the final chapter is not yet written.

When my father's story ended that fateful day when he scared the living daylights out of his airplane qualification instructor, he changed and altered his final life. I guarantee you that man never forgot that adventure, and it puts a smile on my face every time I think of it. My father hit the ground running and left – he stepped out of this life with a smile on his face and a laugh in his heart. What a great way to go like a champion. Together with my mother, he is now resting in Arlington National Cemetery, surrounded by his brothers in arms, in the hallowed halls of American heroes where my father belongs.

18

FOLLOWING HIS FOOTSTEPS

When I look at my life, I hold myself to the highest of standards – what higher or more difficult standard could there be than the example of my father's life? I played high school football, and I know that my father wished I could have been a better player, but sometimes gifts skip a generation. The good news is that when I went to university and I stepped onto that football field, I got a little stronger, and I could see that look in his eyes. Although high school football is pretty cool, college football is a completely different thing. During my medical school years, I participated in rugby and even held my own against a former Rose Bowl champion turned "rugger."

There is only so much you can do with genetics, and I was not quite built for the NFL. Following my father's footsteps and leveraged by sports abilities, my natural athletic prowess led me to the career I always dreamed of, and now I am a doctor. This is an opportunity that would have been nearly impossible when my father was born. He bent the world and created a world in which I was capable of accomplishing my dreams.

Not only did I follow my father's football footsteps, but I also followed in his military footsteps, and in fact, I was commissioned to the Air Force ROTC program.

When I entered the Air Force full time, I was already a medical officer. I had received some financial assistance (a scholarship) from the Air Force that paid for much of my college expenses, particularly my junior and senior years. I thought that the Air Force's quality of life might be better than in the Army – for instance, I knew that the living accommodations for the Air Force were generally a little better. I finished college in 1971 at Saint Louis University, and I took a commission as air force officer in June 1971. My father administered the oath of service to the media at the commissioning ceremony. After that, I went on an educational leave of four years of med school and I did one year of internship, where I did my medical training with minimal Air Force involvement.

In the summer of 1972, I went on active duty for a couple of months and did an externship to get some medical training at the Plattsburgh Air Force Base up in the northeast corner of New York State. I think it is now an international National Guard base as well as a civilian flight port; it is no longer an Air Force base. That was my only time on active duty before coming on the full-time active duty during the summer of 1976. I did a little orientation training down at Sheppard Air Force Base in Texas. Then after a couple of weeks at Shepherd that went up to a place called Thule Air Base in Greenland. I was up there between July 1976 and March 1977.

It was originally going to be a one-year term up there, but for the latter part of the time I was there, the Air Force negotiated a deal with a civilian outfit called the "Danish Arctic Contractors" (DAC), who supplied the English-speaking Danish personnel to staff the hospital. We, the Air Force personnel, then moved on to another assignment. There were a few Air Force contract monitors there, but there were no other medical Air Force medical personnel up there.

From there I went to the Wright-Patterson Air Force Base, which is near my home in Springfield, Ohio. I got to see my parents and live for a while in the old Warrick family home of the mother's side of the family, until I got married in 1978. I lived in Dayton, Ohio for a while until 1981, when I transferred from the active duty Air Force as a Major – one rank higher than my father. I will never forget the proud look in my father's eyes when I caught up and surpassed him. As we get older, we begin to turn into our parents, don't we? That is probably the greatest compliment anyone has ever given me. After leaving active service, I entered the air force reserves where I eventually retired as a full colonel. I wish my father could have been there to see that, but I know he is with me in spirit and a part of him lives on within me.

I lived in Fort Wayne from 1981 to 1988. During that time, I still maintained the affiliation with the Wright-Patterson Medical Center in a reserve capacity. Then in 1990, when the Persian Gulf War broke out, I got assigned to active duty for about seven months down at MacDill Air Force Base at the southern tip of Tampa, Florida. My job there was to do the work of another medical officer who had gone to the Gulf with General Schwarzkopf. Then if something had happened to him over there, I would have gone over to the Gulf as his replacement. I was on active duty for the first Persian Gulf crisis for about seven months.

I got to come home and took another job working in occupational medicine at a Navistar plant in Indianapolis. I did that for a couple of years, and then I took another job while still living in Indianapolis. I worked during the day in an occupational medicine clinic at a Subaru-Isuzu manufacturing plant up in the Lafayette, Indiana, which is about an hour away. I worked there until 2002, and then took more remote jobs, still keeping my house and home in the Indianapolis. I worked during the week just outside of Gary, Indiana, at a steel mill clinic in Burns Harbor, Indiana, and lived during the week in Valparaiso, Indiana, until 2008.

Then I came back to Indianapolis for a company that did urgent care medicine from 2008 until 2014. Since then, I have been working for a company that provides occupational medical services, with some very rare urgent care medicine. That is what I have been doing with my life up to the time of his book.

Every time I help a patient, every time I go to the office, every time

I make a difference, every time my hands heal someone or save a life, I do not do this alone but as an extension of my father and his father, as tools in their hands. Even long after my father's passing, more than thirty years later, he still affects the world every single day, changing and improving lives.

My sister Rosalyn is now a retired auditor with the State Department of Taxation. I know what you are thinking: no one loves to get audited. You have to really love your country to take a job like that, so you can see how me and my sister, in two different ways, continued his legacy of passion for our country, serving in the way that we best can and making the world a better place. I have high hopes for my three children that perhaps they can continue not just my legacy but also my father's legacy and the legacy of his father before him. As we get older, sometimes we begin to notice the continuation of the patterns in our lives, and we realize that legacy is important – it's where we come from. If my father had not accomplished so much, my sister and I would not have had the same opportunities put in front of us.

19

SAYING GOODBYE

Ross Paige Barrett died on January 3, 1985; he was only about sixty-two and a half. He went up with an aviation examiner for a periodic flight competence test, and during that test, he had a cardiac arrest. He had been telling me about some chest discomfort that he had been having. I told him he should see a doctor and get it squared away. I guess maybe he was in some denial in telling us that he had a coronary problem and it might cut short his flying career. I think it would have broken his heart if he could not fly anymore.

When Dad passed away, I was out living and working in Fort Wayne, Indiana. At the time, Dad was still in Springfield, Ohio. He was flying with the Civil Air Patrol and doing some part-time teaching at the high school where he had been the assistant principal before retiring. His official debut retiring was in June of 1982. He went from retired and then some part-time work, a lot of flying, a lot of civil air patrol duties. He continued to officiate a lot of football and basketball games.

When Dad passed away, I was heartbroken. My second son, whom we named Gregory Paige Barrett, had been born seventeen days before, on December 17th, 1984. Dad knew that Greg was on this

Earth, but the two never got a chance to meet. That broke my heart. The other thing I was disappointed in is that after working so hard and going to a couple of wars, Dad was at a point of life where I think he was really looking to enjoy his life and the fruits of his labor. This, from my perspective at the time, was a bitter outcome. I was heartbroken for a long time.

I want to tell you how much I appreciate you coming on this journey with me. I hope that sharing the story of my father will help you find a little bit of inspiration, and I hope you understand why I felt the need to release my soul and share his story. My father was a very private man, he did not tell stories very often, and he was similar to fathers these days in expressing minimal emotions. He was a very quiet and reserved man; he believed that a man spoke with his actions rather than his words, and that is why it has fallen onto me to share his story. I think the only time I saw him cry was at Grandma Warrick's funeral in December 1960. I was too heartbroken to attend Grandma Barrett's funeral, which took place about a year and a half later, but I am sure that he cried then as well.

His actions have already affected so many, and now hopefully my words will magnify the influence and the effect his life has had on the universe. I want everyone to realize that sometimes, a regular person from humble beginnings can make a real difference and have an unbelievably amazing life, achieve dream after dream, and make the impossible possible. I have read somewhere that a truly extraordinary person is one who is indeed extraordinary while appearing quite ordinary.

When you get up tomorrow and think about where your life is headed, I want you to remember one lesson: if you believe hard enough and you push hard enough, just about anything is possible.

FOUND A TYPO?

While every effort goes into ensuring that this book is flawless, it is inevitable that a mistake or two will slip through the cracks.

If you find an error of any kind in this book, please let me know by visiting:

ServeNoMaster.com/typos

I appreciate you taking the time to notify me. This ensures that future readers never have to experience that awful typo. You are making the world a better place.

BOOKS BY JONATHAN GREEN

Serve No Master Series

Serve No Master

Breaking Orbit

20K a Day

Control Your Fate

Habit of Success Series

PROCRASTINATION

Influence and Persuasion

Overcome Depression

Love Yourself

Color Depression Away (coming soon)

Seven Secrets Series

Seven Networking Secrets for Jobseekers